How many baby animals can you name?

How many questions can you answer?
It's fun to peek behind
the flap — to see if you
were right...

What is a baby penguin called?

What is a baby tiger called?

What is a baby

COW

called?

What is a baby
kangaroo
called?

What is a baby

swan called?

What is a baby
cat called?

What is a baby
horse
called?

What is a baby goat called?

What is a baby pig called?

What is a baby
deer called?